What's Under the Sea?

Claudia Diamond

The Rosen Publishing Group's
READING ROOM
Collection™

New York

Published in 2003 by The Rosen Publishing Group, Inc.
29 East 21st Street, New York, NY 10010

Copyright © 2003 by The Rosen Publishing Group, Inc.

First Library Edition 2003

All rights reserved. No part of this book may be reproduced in any form without permission in writing from the publisher, except by a reviewer.

Book Design: Haley Wilson

Photo Credits: Cover, pp. 1, 13 © VCG/FPG International; pp. 2–3, 22–23, 24 © Ron Whitby/FPG International; pp. 4–5 © Peter Gridley/FPG International; p. 7 © Color Box/FPG International; pp. 8, 15, 16 © David Fleetham/FPG International; pp. 10–11, 21 © Telegraph Colour Library/FPG International; p. 15 © Dennie Cody/Ernest Manewal/FPG International; p. 16 © Michael Simpson/FPG International; p. 18 © Carl Roessler/FPG International.

Library of Congress Cataloging-in-Publication Data

Diamond, Claudia C., 1972-
  What's under the sea? / Claudia Diamond.
      p. cm. — (The Rosen Publishing Group's reading room collection)
Includes index.
Summary: Explores the oceans of the world and describes the types of plants and animals that can be found in them.
  ISBN 0-8239-3743-7
  1. Oceanography—Juvenile literature. 2. Marine organisms—Juvenile literature. [1. Oceanography. 2. Marine animals. 3. Marine plants.] I. Title. II. Series.
  GC21.5 .D53 2002
  551.46—dc21
                            2001007484

Manufactured in the United States of America

For More Information
Planet Ocean – DiscoverySchool.com
http://school.discovery.com/schooladventures/planetocean/index.html

Earth's Oceans – EnchantedLearning.com
http://www.enchantedlearning.com/subjects/ocean/

Contents

Earth's Oceans	4
The Ocean Floor	6
Coral Reefs	9
Ocean Plants	11
Fish	12
No Backbone!	14
Crustaceans	17
Ocean Reptiles	19
Ocean Mammals	20
Protecting the Ocean	22
Glossary	23
Index	24

Earth's Oceans

Oceans cover almost three-quarters of Earth's surface, and hold about 97 percent of Earth's water. There are five oceans: the Atlantic, Pacific, Indian, Arctic, and Antarctic. The Pacific Ocean is the largest ocean. It covers almost one-third of Earth's surface!

Most of Earth's living things live in the ocean. The ocean is home to everything from tiny plants and animals to 100-foot-long blue whales.

Ocean water supplies almost all of the moisture that makes rain. Life on Earth couldn't exist without rainfall.

The Ocean Floor

The ocean floor is made of wide areas of flat land, tall mountain chains, **volcanoes**, deep valleys, and **trenches**. When you step into the ocean, you are standing on the **continental shelf**. This slopes down gently until it reaches a drop called the **continental slope**.

The continental slope is much steeper than the shelf. It drops down over two miles to the ocean basin, which is mostly made up of flat land.

> It might be hard to imagine, but there are slopes, mountains, trenches, and wide plains below the ocean's surface.

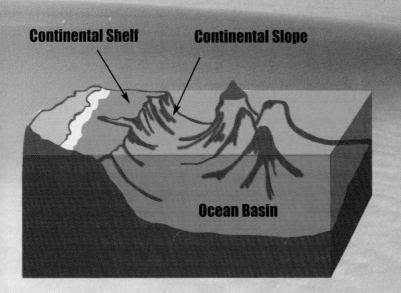

Coral Reefs

Many kinds of ocean creatures live in **coral reefs**. Coral reefs are made of millions of tiny coral animals. Coral animals live in warm, shallow salt water. When coral animals die, other coral animals attach themselves to their skeletons. This is how a coral reef is formed.

The Great Barrier Reef off the coast of Australia is the world's longest coral reef. It is about 1,250 miles long!

> Millions of living coral animals are what give this coral reef its bright, beautiful colors.

Ocean Plants

Many kinds of plants grow in the ocean. They grow close to the ocean's surface where sunlight can reach them. Plants need sunlight to grow. There are no plants in deeper parts of the ocean because there is no sunlight.

Some ocean plants are very small. Other plants, like seaweed and sea grasses, are much larger. Large, brown seaweed called giant kelp can grow up to 200 feet long and can form large underwater forests. Kelp is used to make some kinds of ice cream!

Plants can grow as deep as 330 feet below the ocean's surface. In deeper water, there is not enough light for plants to grow.

Fish

Over 13,000 kinds of fish live at all different levels of the ocean. More than a third of these live in the waters around coral reefs. Fish that live around coral reefs are usually the most colorful.

Fish found on the ocean's floor are sometimes very unusual looking. Because there isn't much food to eat deep in the ocean, these fish often have big mouths and big stomachs that hold a lot of food at once. They eat dead plants and animals that float down from the surface.

> Fish that live around coral reefs use their beautiful colors and shapes to attract mates.

No Backbone!

Most animals that live in the ocean are **invertebrates** (in-VER-tuh-brayts). Invertebrates are animals without backbones.

Animals called **mollusks** make up the largest group of ocean invertebrates. Mollusks don't have any bones at all! Clams, oysters, snails, octopuses, and squids are all mollusks. Mollusks have soft bodies, but most have hard outer shells to protect them from enemies. Octopuses and squids don't have outer shells.

> These animals may look very different, but they are all mollusks.

Crustaceans

Crustaceans (krus-TAY-shuns) are also invertebrates. Like mollusks, crustaceans have no bones. They have hard outer shells that protect their bodies. They also have feelers and many jointed legs. Crabs, shrimp, and lobsters are all crustaceans.

Crustaceans are an important part of the ocean food chain. Smaller crustaceans eat tiny plants. Bigger crustaceans, fish, and whales eat the smaller crustaceans.

> There are more than 42,000 different kinds of crustaceans. They live in salt water, in fresh water, and on land.

Ocean Reptiles

Some kinds of **reptiles** also live in the ocean. Ocean reptiles include sea turtles, sea snakes, and some lizards. Unlike fish, mollusks, and crustaceans—which get their oxygen from water—reptiles need air to breathe. They store air in their lungs and can stay underwater much longer than humans can.

Most reptiles hatch from eggs on land and then make their way to the ocean. Reptiles often live in warm areas where they can lie in the sun to warm their bodies.

> Some sea turtles can grow to be about 8 feet long and can weigh up to 1,500 pounds!

Ocean Mammals

Seals, walruses, dolphins, and whales are **mammals** that live in the ocean. Like ocean reptiles, ocean mammals also need air to breathe. Seals and walruses can stay underwater for many minutes using air stored in their lungs. Dolphins and whales can also store air in their lungs. They breathe through **blowholes** in the tops of their heads. Ocean mammals have a layer of fat to keep them warm. They give birth to live babies and feed them milk from their bodies.

> Dolphins play and hunt for food together. They talk to each other by making whistling and clicking sounds.

21

Protecting the Ocean

Pollution puts the oceans' plants and animals in danger every day. Dolphins and other ocean animals are killed when they get caught in fishing nets left in the water. Oil spills and **chemicals** dumped into the oceans kill many plants and animals. Now there are laws to keep people from dumping things into the oceans.

We can help protect the oceans by cleaning up after ourselves when we go to the beach. All life on Earth depends on the oceans, so we must work harder to take care of them. We have much more to learn about all the life inside our oceans.

Glossary

blowhole — A hole for breathing on the tops of some ocean mammals' heads.

chemical — A man-made substance that can be dangerous if not handled properly.

continental shelf — The gentle slope of the ocean floor that begins where the land meets the water.

continental slope — A steep drop from the continental shelf to the bottom of the ocean.

coral reef — A chain of coral where many ocean creatures live.

crustacean — An animal with a hard shell, jointed limbs, feelers, and no backbone.

invertebrate — An animal without a backbone.

mammal — A warm-blooded animal with a backbone. Females have live babies and feed them milk from their bodies.

mollusk — An animal with a soft body and no bones that is usually covered by a hard shell.

pollution — Anything that makes Earth unclean.

reptile — A cold-blooded animal that usually hatches from an egg and has scales, such as a turtle, a lizard, or a snake.

trench — A narrow ditch with steep sides.

volcano — An opening in Earth's crust through which hot liquid rock is sometimes forced.

Index

A
air, 19, 20
animals, 5, 9, 12, 14, 22

B
bones, 14, 17

C
continental shelf, 6
continental slope, 6
coral reef(s), 9, 12
crustaceans, 17, 19

E
Earth, 4, 5, 22

F
fish, 12, 17, 19
food, 12, 17

I
invertebrates, 14, 17

M
mammals, 20
mollusks, 14, 17, 19

O
ocean basin, 6

P
plants, 5, 11, 12, 17, 22

R
reptiles, 19, 20

S
shells, 14, 17
sunlight, 11

W
water(s), 4, 9, 12, 19, 22
whales, 5, 17, 20